Annals of the ICRP

ICRP PUBLICATION 87

Managing Patient Dose in Computed Tomography

Editor
J. VALENTIN

PUBLISHED FOR

The International Commission on Radiological Protection

UK	Elsevier Science Ltd, The Boulevard, Langford Lane, Kidlington, Oxford OX5 1GB, UK
USA	Elsevier Science Inc., 660 White Plains Road, Tarrytown, New York 10591-5153, USA
JAPAN	Elsevier Science Japan, Tsunashima Building Annex, 3-20-12 Yushima, Bunkyo-ku, Tokyo 113, Japan

---

Fourth edition 2000
ISBN 008 044 0835
ISSN 0146-6453

Published quarterly (March, June, September, December)
December 2000 issue

Typeset by Variorum Publishing Ltd, Rugby
Printed and bound in Great Britain by Polestar Wheatons Ltd, Exeter

ICRP Publication 87

Annals of the ICRP

Editorial

DON'T STEP ON IT

Computed tomography (CT) is a wonderful tool in medical practice, and in many cases a life-saving resource when rapid decisions are needed in the emergency room. Like many other tools, it is admirable as a brilliant feat of engineering and possesses a kind of beauty in its own right; like many other tools, if misused it can be dangerous to all and sundry in its vicinity.

This report addresses the management of patient dose in computed tomography. Its purpose is not in any way to impede or question the increasing use of computed tomography as such. To continue the tools parable, if a tool is useful, we do not want to stop people from using it. However, we may wish for designers and manufacturers of that tool to keep possible risks in mind and perhaps include safety features that will help users to avoid unnecessary risks. Furthermore, we may want to make sure that users are aware of potential risks in over enthusiastic use of the tool and utilise it in a safe way.

This is essentially what the present report aims to do for you: ICRP recommends that manufacturers of CT equipment should consider safety features that could reduce the workload for users (in particular, automatic exposure control), and ICRP recommends that users ask themselves two simple questions: Is CT the right tool to use in this case, and am I using the CT tool in the right way. In other words, the classical ICRP tenets, justification of the procedure and optimisation of radiological protection. The details in the report are of course adapted to the specific technical features of computed tomography, but the underlying philosophy is the same as in any other radiological protection exercise in diagnostic medicine.

Avid readers of the *Annals of the ICRP* will have observed that this report is just one of a whole series of topically focused publications that ICRP is currently developing in the area of radiological protection in medicine. This is where ICRP originated; we were created back in 1928 by the 2nd International Congress of Radiology and are still linked to the International Society of Radiology. However, ICRP of course deals with many issues outside the medical area too, and in the foreseeable future, we are likely to continue to issue publications both concerning medical uses of radiation and other, often completely different, areas of radiological protection.

So, you may ask, why is there still such an interest in radiological protection in medicine? For one thing, the numbers of procedures are huge. Based on information from the United Nations Scientific Committee on the Effects of Atomic Radiation, world-wide, there are some 2 billion diagnostic x-ray examinations annually, about 32 million nuclear medicine procedures, and some 5.5 million radiotherapy treatments. The doses and attendant risks run from vanishingly small to very big (and CT doses are not vanishingly small, particularly since many patients are subjected to several successive CT examinations). Technology and methods are developing at a furious pace, to the great benefit of patients but at times perhaps leaving protection a step or two behind. Some newer modalities, like digital radiology, should in theory permit lower average doses to patients but have in fact led to the opposite, due to incorrect use of the equipment.

This in turn reflects one of the root causes behind what ICRP perceives as a need for renewed and reinforced activities in radiological protection in medicine: the effects of radiation and protection against such effects no longer feature prominently on the syllabus of the average medical student. Therefore, ICRP intends to produce some material that could be helpful in education and training (watch this space and our web site, www.icrp.org, for more information). And therefore, we are making recommendations intended to help medical staff to avoid unnecessary risks. For instance, fluoroscopic equipment is now often fitted with a foot pedal that operates an option with an alluring name like 'boost' or 'turbo'. Like the accelerator of a car, this is most useful but can easily be misused. Like we all want young drivers use their accelerators sensibly, we would also like operators of fluoroscopic equipment to tell themselves: 'Don't step on it–unless it's really needed'.

Jack Valentin

CONTENTS

PREFACE

Over the years, the International Commission on Radiological Protection (ICRP), referred to below as 'the Commission', has issued many reports providing advice on radiological protection and safety in medicine. Its *Publication 73* is a general overview of this area. These reports summarise the general principles of radiation protection and provide advice on the application of those principles to the various uses of ionising radiation in medicine and biomedical research.

Most of these reports are of a general nature, and the Commission wishes to address some specific situations where difficulties have been observed. It is desirable that reports on such problem areas be written in a style which is accessible to those who may be directly concerned in their daily work, and that every effort is taken to ensure wide circulation of such reports.

A first step in this direction was taken at the Commission's meeting in Oxford, United Kingdom, in September 1997. At that time, on the recommendation of ICRP Committee 3, the Commission established several Task Groups to produce reports on topical issues in medical radiation protection.

Some such reports have now appeared in print, and the present one is the result of the work of yet another one of those Task Groups. The terms of reference of this Task Group on Control of Radiation Dose in Computed Tomography were to prepare a report highlighting the increasing frequency and radiation doses in computed tomography and documenting the steps necessary to manage the radiation dose. The Task Group was requested to address the various levels that need to be involved in dose management, including those of manufacturer, operator, radiologist, and referring physician.

The membership of the Task Group was as follows:

M.M. Rehani (Chairman)	G. Bongartz	S.J. Golding
L. Gordon	W. Kalender	T. Murakami
P. Shrimpton		

Corresponding members were:

R. Albrecht	K. Wei

Numerous helpful comments were received from interested parties through the Commission's customary public consultation via the Internet.

The membership of Committee 3 during the period of preparation of this report was:

F.A. Mettler, Jr. (Chairman)	J.-M. Cosset	M.J. Guiberteau
L.K. Harding (Secretary)	J. Liniecki (Vice-Chairman)	S. Mattsson
H. Nakamura	P. Ortiz-Lopez	L.V. Pinillos-Ashton

M.M. Rehani	H. Ringertz	M. Rosenstein
Y. Sasaki	C. Sharp	W. Yin
W.Y. Ussov		

This report aims to serve the purposes described above. In order to be as useful as possible for those purposes, its style differs in a few respects from the usual style of the Commission's publications in the *Annals of the ICRP*.

The report was approved for publication by the Commission in October 2000.

Annals of the ICRP

Managing patient dose in computed tomography

ICRP Publication 87

Approved by the Commission in October 2000

Abstract-Computed tomography (CT) examinations can involve relatively high doses to patients. The doses can often approach or exceed levels known with certainty to increase the probability of cancer. The frequency of CT examinations is increasing worldwide and the variety of examinations is also increasing. However, in contrast to the common trend in diagnostic radiology, the rapid developments in CT have not led in general to a reduction of patient doses per examination. Therefore, management of patient dose is crucial. Proper justification of examinations, use of the appropriate technical parameters during examinations, proper quality control, and application of diagnostic reference levels of dose as appropriate would all contribute to this end. There is also scope for further technical development of the equipment used. The present publication aims to provide information in all these respects in order to provide assistance in the successful management of patient dose.

Keywords: CT; Absorbed dose; Radiation protection; Diagnostic radiology; Dose reduction

MAIN POINTS

- **The absorbed doses to tissues from computed tomography (10–100 mGy) can often approach or exceed the levels known to increase the probability of cancer.**
- **CT examinations are increasing in frequency.**
- **Technical and clinical developments in CT have not led in general to reductions in patient dose per examination, in contrast to the trend for conventional radiology.**
- **Referring physicians and radiologists should make sure that each CT examination is clinically indicated.**
- **Radiologists are responsible for managing the dose in collaboration with operators and medical physicists. Operators should be trained and skilled in optimising CT examinations.**
- **Many practical possibilities currently exist to manage dose. The most important one is to reduce mAs and scan volume.**
- **Paediatric patients should have specific protocols with lower exposure factors (especially mAs).**
- **Automatic exposure control would be the most helpful improvement in CT equipment for dose management.**

SUMMARY

1. What is the motivation for this report?

The motivation for this report stems from the relatively high radiation doses to the patient in CT examinations and the increasing frequency and variety of examinations. This report aims to provide radiologists and other clinical and technical staff with the information to manage successfully patient doses.

2. How high are the doses?

Absorbed doses in tissues from CT are among the highest observed in diagnostic radiology (i.e. 10-100 mGy). In view of a tendency to repeat the CT examinations, these doses can often approach or exceed levels at which an increased cancer incidence has been directly observed in human populations.

3. What practical actions can be used to manage patient dose?

The referring physician should evaluate whether the result of each examination will affect patient management. The radiologist should be satisfied that the procedure is justified. The operator[1] should be aware of the possibilities to reduce patient doses by adapting technical parameters to each patient and the examination at hand, with special attention being paid to paediatric and young patients. More than a 50 percent reduction in patient dose is possible by an appropriate choice of technical parameters, attention to quality control, and the application of diagnostic reference levels in co-operation with a medical physicist.

4. What new equipment features would help manage patient dose?

CT doses are relatively high and have not decreased over time as they have in conventional radiography. Further improvements in CT equipment could help the operator to reduce unnecessary patient doses substantially. The most important of these features will be anatomically based on-line adjustment of exposure factors and new image reconstruction approaches associated with multi-slice computed tomography.

---

[1] The nomenclature differs in different countries; depending on the location, the person operating the equipment during an examination may be called a radiographer, a radiological technologist, a medical imaging technologist, a member of the imaging staff, or some other term. In this report, the term 'operator' is used throughout.

1. WHAT IS THE MOTIVATION FOR THIS REPORT?

(1) The motivation for this report stems from the relatively high radiation dose to the patient in CT examinations and the increasing frequency and variety of examinations. This report aims to provide radiologists and other clinical and technical staff with the information to manage successfully the patient doses.

1.1. General introduction

(2) Computed tomography (CT) was introduced into medical imaging in 1972 and since then has evolved rapidly in terms of both technical performance and clinical use. Although initial experiences readily predicted widespread implementation of the technique, it could hardly have been foreseen how rapidly CT would become one of the most important type of x-ray procedure worldwide. Spiral CT and in particular the latest-generation of scanners with multi-slice capability in subsecond time frames have allowed improvements in the speed of acquisition and the quality of images. This has resulted in highly reliable information about every part of the body, without motion artefacts from peristalsis and breathing. The consequence has been further growth of the modality. Thus, completely new indications for CT are being reported, as well as completely new methods for performing and reading the studies. Twenty years ago, a standard CT examination of the thorax took several minutes to conduct, while today similar information can be accumulated within a single breath-hold period. This makes it more comfortable for patients and also easier for physicians to refer patients for examination, since the investigation is fast, well tolerated, accessible and, last but not least, regarded as highly reliable in its outcome.

(3) Shortening data acquisition time does not necessarily lead to a reduction in patient radiation dose since investigations are then also more easily extended or repeated.

(4) The benefit to patients from properly directed CT scanning is beyond any doubt in relation to an extensive list of indications, from the use of cranial CT in unclear neurological status to orthopaedic investigations, as a preparation to the planning of surgical procedures. Various therapeutic regimens are directed by the results of CT investigations for staging disease and treatment planning in oncology patients. In modern medicine, where the evaluation of cost effectiveness plays a dominant role, an expensive examination like CT scanning can save money by excluding patients from inappropriate and even more expensive therapeutic procedures. Patient management has changed in relation to a multitude of emergency cases as well as in diagnostic routine, with CT often being the initial investigation in a diagnostic work-up. Some clinical investigations are postponed until after the results from CT scanning have been obtained in order to save time and money.

(5) In view of the known risk of radiation-induced cancer at dose levels reachable by CT examinations, there is a continuing need to balance the benefits and risks to patients. In principle, this means the elimination of unnecessary exposure. In practice, it requires the prior clinical justification of all CT examinations so as to ensure a net positive benefit for each patient. Justification should be followed by the adop-

tion of imaging techniques that will maximise the benefit relative to harm, by keeping the patient dose as low as reasonably practicable to meet defined clinical needs (*Publication 73*, ICRP, 1996).

(6) In the early 1990s, when magnetic resonance imaging (MRI) had emerged and safety considerations concerning radiation exposure from medical imaging were gaining special importance in the western world, CT applications reached a short plateau and were expected to decline. However the pendulum has recently swung back towards CT due to technical innovations resulting in ultra-fast CT. Increasing numbers of CT scanners per million population observed in the last decade are the consequence of the high acceptance of this method in clinical imaging. Three-dimensional display of body regions like the abdomen, chest, intracranial, and osseous structures is very useful. Online reconstructions in coronal, sagittal or any oblique plane help clinicians to understand better the underlying pathoanatomy of a disease.

(7) Greater implementation of CT in routine patient management has increased the radiation burden from CT. In addition, requirements for improved image quality have led to higher patient doses. While new low-dose CT examinations are being proposed for a number of screening investigations, such as for lung cancer, patient doses have risen overall due to the steadily increasing demands for adequate spatial and contrast resolution in the majority of diagnostic CT examinations. In conventional CT scanning, patient exposure is restricted to a thin slice of the body during each rotation of the x-ray tube with the possibility of an inter-slice gap. However, in spiral CT and more so in multi-slice CT, the cumulative radiation dose from each complete investigation can be relatively high, unless pitch greater than 1 is chosen, and gives rise to concern. Accordingly, CT is specifically addressed by several guidelines or regulations, mainly in the European Union (CEC, 1997; 1999). The latest Euratom Directive on medical exposures (CEC, 1997) classified CT, together with interventional radiology, as a high-dose radiological procedure.

(8) Substantial variations in the dose to the patient for similar types of CT examination can result when applying different imaging protocols or using different CT scanners. Even when considering an identical clinical problem, the imaging procedure employed at two different imaging centres may be completely different. Comparison of the final diagnosis is not a measure for quality assessment or dose comparison. Up to a certain point, image quality usually increases with radiation dose. Past a certain level, the dose continues to increase but the impact of the procedure on patient care may remain the same. Among the factors which may be difficult to substantiate scientifically, but are important in practice like diagnostic uncertainty, ‘more-is-better’ philosophy, or even competition, can produce CT protocols that are too extensive, too long, and repetitive. Repeat investigations and examinations in paediatric patients and during pregnancy must be given with special care. The problem of high radiation dose from CT needs to be acknowledged, guidelines for quality control and CT dose measurements need to be promulgated, and there needs to be optimisation of CT with respect to radiation protection.

1.2. Statistics

(9) In a 1989 survey, the UK National Radiological Protection Board, NRPB, reported that 20% of the national collective dose from all medical x-ray examinations was derived from CT alone, although it represented only 2% in terms of the total number of such examinations (Shrimpton et al., 1991). Following further increases in the number of scanners available in the UK and growing implementation of the technique, subsequent reviews have suggested that the contribution from CT to collective dose had risen to approximately one third in 1995 and about 40% in 1998 (Rehani and Berry, 2000; Shrimpton and Wall, 1995; Shrimpton and Edyvean, 1998; Wall and Hart, 1997).

(10) Worldwide, CT constitutes 5% of all medical x-ray examinations and provides a contribution of 34% to the resultant collective dose (Fig. 1) according to the United Nations Scientific Committee on the Effects of Atomic Radiation (UNSCEAR, 2000). In those countries with the highest levels of health care (classified by UNSCEAR as level I countries), the corresponding contributions are 6% and 41%, respectively. These data have increased relative to comparative analyses for frequency and collective dose for the period 1985- 1990: 3% and 14%, respectively, for the entire world, and 4% and 18%, respectively, for industrialised countries.

(11) Such trends should of course be regarded not only in the light of increasing CT application, but also with regard to a decline in conventional x-ray exposures following initiatives for optimising the protection of patients. The worldwide total number of CT scanners is approximately 34,000 (UNSCEAR, 2000), with 80% operating in the western world where one quarter of the world's population lives. Annually, there is a global total of about 93 million CT examinations, corresponding to a frequency of 16 per 1000 inhabitants worldwide; 90% of all these procedures are conducted in the western world (UNSCEAR level I countries), at a rate of 57 examinations per 1000 population, with about 6% involving children (0-15 years). Although there is a remarkable variation in the numbers of CT scanners and investigations between different countries, the tendency towards increasing collective radiation dose is similar.

1.3. Trends

(12) The application of CT in clinical routine has greatly increased in frequency in recent years due to technical advances in equipment leading to much faster image acquisition and processing capability. A recent paper from the USA (Mettler, 2000) indicates that in a department with a typical referral pattern, CT scans now constitute 11% of all x-ray examinations and contribute 67% to the collective dose. Eleven percent of CT was performed on children below the age of 15 and most CT patients had more than one scan sequence on the same day. The contribution from CT examinations to population exposure has steadily increased over the years. In radiological protection terms, the significance of this lies not merely in the absolute

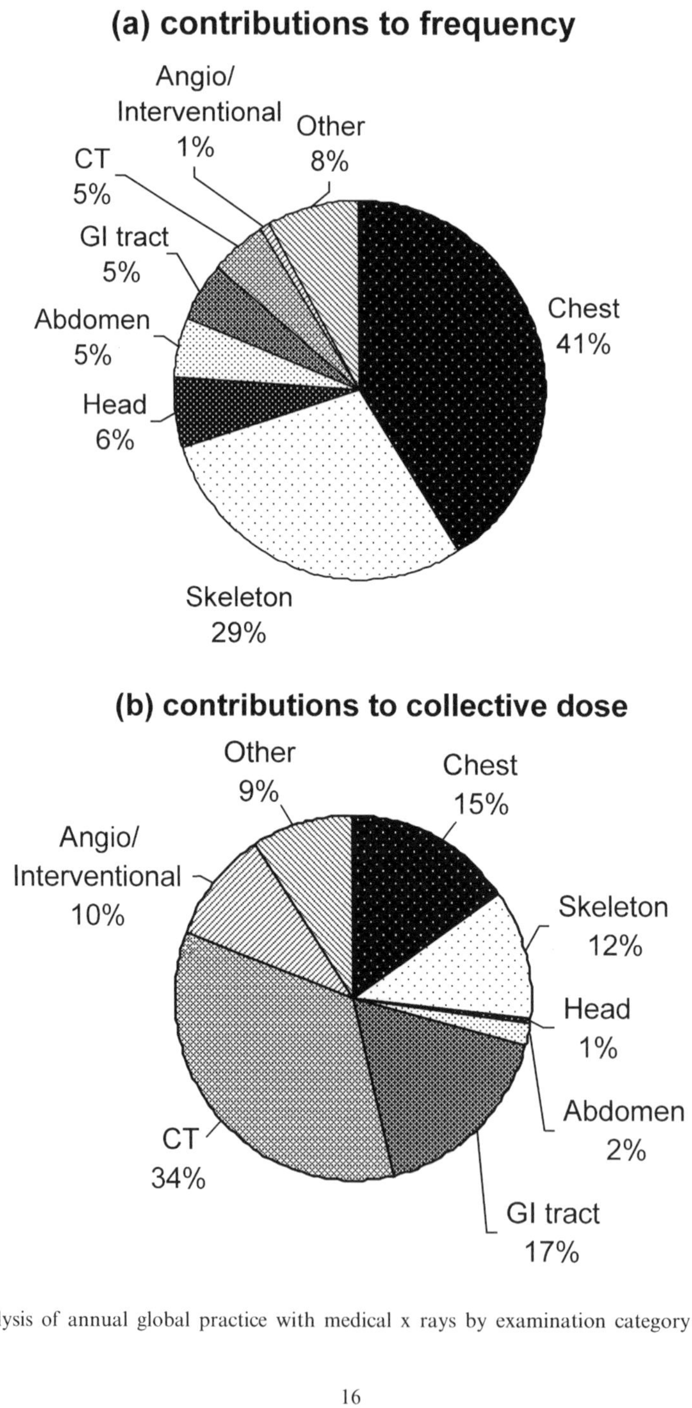

Fig. 1. Analysis of annual global practice with medical x rays by examination category (UNSCEAR, 2000).

magnitude of the collective dose, but more in the increased potential for unnecessary patient exposures and the potential for dose reduction.

(13) New opportunities for dynamic and continuous imaging provided by CT fluoroscopy and angiography have given impetus to practice in interventional radiology. Even though clinically useful, these developments are contributing to larger radiation dose to the patient. Interventional procedures using CT expose specific parts of the body for an extended period of time, thus adding to localised dose. In CT fluoroscopy, staff exposures are increased since they have to be present inside the CT room, near the gantry, and their hands may be in the primary beam.

1.4. Objectives of this report

(14) This report gives advice on managing the problem of increasing radiation doses from CT by technical means, training, and the justification of individual examinations. It is addressed to radiologists, operators[2], physicists, and referring physicians, as well as to manufacturers, professional bodies, and national authorities.

[2] The nomenclature differs in different countries; depending on the location, the person operating the equipment during an examination may be called a radiographer, a radiological technologist, a medical imaging technologist, a member of the imaging staff, or some other term. In this report, the term 'operator' is used throughout.

2. HOW HIGH ARE THE DOSES?

(15) Absorbed doses in tissues from CT are among the highest observed in diagnostic radiology (i.e. 10–100 mGy). In view of a tendency to repeat the CT examinations, these doses can often approach or exceed levels at which an increased cancer incidence has been directly observed in human populations.

2.1. Introduction

(16) The conditions of exposure in CT, in which thin slices of the patient are irradiated in rotational geometry by a fan beam of x-rays, are quite different from those in conventional x-ray examinations. Therefore, specific techniques of dosimetry have had to be developed both to assess patient doses and to allow monitoring of performance for different types of CT examinations. Patient dose should not, of course, be considered in isolation from image quality (CEC, 1999). The quantities used in this text are summarised below:

- *Absorbed dose in tissue:* Energy deposited in tissue/organ per unit mass, measured in Gy (gray). The basic quantity used for assessing the radiation risk to the tissue/organ.
- *Effective dose:* A calculated quantity that takes into account the difference in radiosensitivity of tissues. It is used as an index to compare relative radiation risk from different radiological procedures and is expressed in Sv (sievert).
- *Collective dose:* The sum of effective doses in a patient population. Measured in man-Sv.
- *$CTDI_w$* and *DLP:* Computed Tomography Dose Index (weighted) and Dose Length Product, respectively. These relate to measurements in standard phantoms and represent dosimetric quantities for determining relative performance of the equipment and technique used in CT. $CTDI_w$ is measured in mGy and DLP in mGy cm. Either quantity can be used for the setting of diagnostic reference levels.

2.2. Which quantities should be used to assess patient dose?

(17) X-ray exposures of patients are most completely characterised by the absorbed radiation doses to each organ or tissue of the body (UNSCEAR, 2000), although this approach is rather unwieldy and difficult for routine use. The weighted summation of organ doses to yield the quantity effective dose (ICRP, 1991) provides a convenient index of overall exposure that is useful for broad comparison of different CT techniques and other types of radiological examination.

(18) Since the direct measurement of absorbed dose is impractical for most organs, comprehensive dose assessment necessarily involves the simulation of clinical CT practice, utilising physical or mathematical representations of the patient (anthropomorphic phantoms). Distributions of absorbed dose in such phantoms may be determined either by measurements (Mini et al., 1995; Nishizawa et al., 1995) or,

with greater utility, by computational modelling (Zankl, 1998). The latter approach has provided dose coefficients, normalised to a free-in-air axial dose, that allow the estimation of organ and effective doses in standard adult (Jones and Shrimpton, 1993; Shrimpton and Edyvean, 1998; Zankl et al., 1991; Kalender, Wolf, Suess et al., 1999) and paediatric patients (Caon et al., 2000; Zankl et al., 1993; Zankl et al., 1995) for particular scanning protocols.

(19) Building on initial experience with geometrical mathematical phantoms, computational methods of dosimetry are advancing steadily with the development of more realistic (voxel) phantoms based on digital images of humans (Veit et al., 1989; Caon et al., 1999; Jones, 1997; Xu et al., 2000). Differences in the results from calculations for different anthropomorphic phantoms under similar conditions of exposure underline the limitations and uncertainties in such computed dose coefficients. Accordingly, results determined for standard phantoms should not be applied to examinations of individual patients (Zankl, 1998), although patient-specific computational modelling is also becoming a reality. In general, there is also reasonable agreement between sets of organ doses derived from measurements or calculations for a given CT examination technique when account is taken of differences in the exposure conditions being modelled (Calzado et al., 1995; Geleijns et al., 1994; Seifert et al., 1995). Some very general values of organ and effective dose for common types of CT and other x-ray examinations are shown in Tables 1 and 2.

2.3. Which quantities should be used to monitor performance?

(20) Notwithstanding the need for some assessment of effective dose, there is also a requirement for periodic monitoring using simpler, more-easily measured dose quantities to characterise local performance, as an essential part of routine quality assurance. The quantities commonly used for monitoring patient doses during conventional x-ray examinations, such as entrance surface dose, are less useful in CT. In practice, dosimetry in CT is based principally on measurements of the computed tomography dose index (CTDI), most conveniently quantified by using a pencil ionisation chamber with an active length of 100 mm and calibrated in terms of absorbed dose to air (see Appendix A).

Table 1. Typical doses during CT on adults (Shrimpton et al., 1991)

CT Examination	Eyes (mGy)	Thyroid (mGy)	Breast (mGy)	Uterus (mGy)	Ovaries (mGy)	Testes (mGy)	Effective dose (mSv)
Head	50	1.9	0.03	*[a]	*	*	1.8
Cervical spine	0.62	44	0.09	*	*	*	2.6
Thoracic spine	0.04	0.46	28	0.02	0.02	*	4.9
Chest	0.14	2.3	21	0.06	0.08	*	7.8
Abdomen	*	0.05	0.72	8.0	8.0	0.70	7.6
Lumbar spine	*	0.01	0.13	2.4	2.7	0.06	3.3
Pelvis	*	*	0.03	26	23	1.7	7.1

[a] The symbol * indicates that the dose is <0.005 mGy.

Table 2. Comparison of typical doses in UK from CT and conventional x-ray examinations (RCR, 1998)

Diagnostic procedure	Typical effective dose (mSv)
Conventional x-ray procedure	
Limbs and joints	<0.01
Chest (single PA film)	0.02
Skull	0.07
Thoracic spine	0.7
Lumbar spine	1.3
Hip	0.3
Pelvis	0.7
Abdomen	1.0
IVU	2.5
Barium swallow	1.5
Barium meal	3
Barium follow through	3
Barium enema	7
CT	
Head	2
Chest	8
Abdomen	10
Pelvis	10

(21) When such CTDI measurements are made free-in-air on the axis of rotation, they provide by themselves only a coarse indication of patient exposure, although they can be used with dose coefficients from mathematical modelling to estimate organ and effective doses for particular scanning techniques (Shrimpton and Wall, 1995). However, measurements of CTDI made in a phantom are able to better reflect the influence of scanner design, particularly the use of beam shaping filters, upon the radial distribution of absorbed dose within the irradiated slice. In particular, a weighted sum of CTDI measurements ($CTDI_w$) made at the centre and 10 mm below the surface of standard CT head and body dosimetry phantoms provides a useful way of characterising exposure in CT (CEC, 1999); see Appendix A. Values of $CTDI_w$ are recommended for display on the operator's console of the CT scanner (IEC, 1999), reflecting the parameters of operation selected. However, operators should understand the basis of the values displayed, particularly where these include a correction for the pitch, that is, table feed in one rotation relative to collimation ($CTDI_{w,\ eff}$) (Nagel, 2000; IEC, 1999).

(22) Whereas it is inappropriate to impose limits on the doses received by patients for medical purposes, the concept of diagnostic reference levels (DRLs) is increasingly recognised as a useful way of promoting the fundamental requirement for optimisation of patient protection (ICRP, 1996). Diagnostic reference levels are set in terms of practical dose quantities that characterise clinical practice and so allow comparisons within and between clinical facilities. Such dose levels can be established for different types of examination on the basis of wide-scale survey data in order to help identify potentially inadequate performance (Wall and Shrimpton,

1998). This approach has already proved effective for reducing unnecessary irradiation from conventional x-ray examinations (Hart et al., 1996).

(23) Two reference dose quantities have been defined for the purpose of promoting the use of good technique in CT: (1) weighted CTDI ($CTDI_w$) per rotation and (2) dose-length product (DLP) which takes into account beam collimation and the number of rotations in a complete examination (CEC, 1999); see Appendix A. These dose quantities relate to measurements in standard head or body dosimetry phantoms, for a specific type of examination and the exposure conditions used in clinical practice. The concept was initially developed for examinations on adult patients, although it has subsequently been extended to paediatric CT (Shrimpton and Wall, 2000). Monitoring of $CTDI_w$ per rotation takes account of the exposure settings selected, such as tube current and tube voltage. Monitoring of DLP for a complete examination takes account also of the volume of irradiation, as determined, for example, by the number of slices in serial scanning or the acquisition time in spiral scanning, and the number of such scan sequences conducted during the examination. Values of DLP may also be used to derive broad estimates of effective dose for CT procedures using region-specific coefficients (CEC, 1999; Shrimpton and Wall, 2000).

(24) In line with international recommendations, some initial diagnostic reference levels for CT have been proposed for common examinations on adults (CEC, 1999) and paediatric patients (Shrimpton and Wall, 2000). Such levels are for comparison locally in CT facilities against the measured values of dose descriptors assessed during examinations on representative groups of patients and they should not be applied on an individual patient basis. Diagnostic reference levels are intended to act as guides that trigger internal investigations by departments to identify situations where improvements in dose management may be necessary. Their use is discussed further in Chapter 3.

2.4. What influences the patient dose?

(25) Patient dose in CT is determined by the inherent characteristics of the scanner, the size of the patient, the anatomical region under investigation, and the scanning protocol and technique used. The absorbed dose should be sufficient to meet the particular clinical need. These issues are discussed more fully in Chapter 3. The influence of changes in some key technical and operational parameters on absorbed dose in tissues are summarised in Table 3 (Kalender, 2000); these apply both for serial or spiral scanning. Patient dose (both organ and effective doses) depends strongly on the radiation quality of the x-ray beam. Patient dose decreases, for a given level of image quality (and in particular noise), with increasing tube voltage or filtration.

(26) For a given scanner, dose is linearly related to the product of tube current (mA) and examination time (s). The absorbed dose for a given body part will also depend inversely on the pitch (table travel per rotation relative to beam collimation), for constant mA, during serial or spiral scanning. Comparison of mAs values for different models of scanners is unlikely to provide meaningful information on relative dose due to differences in their design. Finally, the number of organs exposed

Table 3. Influence of technical and operational parameters on patient dose during CT (Kalender, 2000)

Parameter	[Influence on patient dose]
High tube voltage	Higher kV advantageous (for constant image noise)
Filtration	Higher filtration advantageous
Tube current	Linear increase with mA
Scanning time	Linear increase with s
Slice thickness	Approximately linear increase in dose with thickness (valid for single slices)
Scan volume	Approximately linear increase in dose with volume

and their mean doses depend on the volume of the patient irradiated during an examination. These will both increase with the number of slices in serial scanning or the acquisition time in spiral scanning and the number of such sequences performed during a complete examination (for example multi-phase contrast scans of the liver). The absorbed dose for a given body part will also depend inversely on the pitch (table travel per rotation relative to beam collimation) during serial or spiral scanning. Scan projection radiography, which is commonly conducted to aid localisation in CT scanning, typically contributes only a few percent of the total patient dose (Shrimpton et al., 1991; Mini et al., 1995). New technical developments for reducing patient dose, such as tube current modulation, are discussed in Chapter 4.

2.5. What are the typical levels of patient dose?

(27) Patient doses in CT are typically higher than those associated with many other common types of diagnostic x-ray procedure. Some illustrative doses to selected organs are shown in Table 1, on the basis of mean values reported in one particular comprehensive survey of national practice (Shrimpton et al, 1991); these may not be representative of practice in all countries and other studies of organ doses in CT have been reviewed elsewhere (UNSCEAR, 2000), although comparable national data remain sparse. Doses to individual patients may be significantly higher than such mean data. For example, uterine absorbed doses of up to 80 mGy have been reported during pelvic CT and particular care is therefore required when conducting such examinations on female patients of reproductive age in order to avoid unnecessary fetal exposures (Sharp et al., 1998). The dose to the thyroid, breast, and testes from scattered radiation may be further reduced by the use of lead shielding (Beaconsfield et al., 1998; Hidajat et al., 1996; Price et al., 1999). Lower levels of patient dose are often possible in CT with attention to choice of scanning technique, particularly with regard to lower settings or dynamic modulation of tube current (UNSCEAR, 2000); see Chapters 3 and 4.

(28) Typical effective doses to adults from some routine CT and conventional diagnostic x ray examinations in the UK are shown in Table 2 (RCR, 1998). Such dose data are broadly comparable with practice reported in other countries (UNSCEAR, 2000). Effective doses in CT are in general relatively high (typically 1–30 mSv) and may be similar to those values observed for some complex angio-

graphic and interventional radiological procedures (UNSCEAR, 2000). Values of weighted CTDI and DLP are typically in the ranges 10–100 mGy and 50–2000 mGy cm, respectively (CEC, 1999).

(29) With the use of standard scanning techniques, the energy imparted to the patient in CT increases with patient size, although the calculated effective dose is somewhat higher for children than adults; for example, data from one particular institution indicated values of 6.0 mSv (newborn) and 1.5 mSv (adult) during head examinations, and 5.3 mSv (newborn) and 3.1 mSv (adult) during abdomen examinations (Huda et al., 1997).

(30) Typical values of patient dose in CT can be expected to change with developments in technology (spiral, multi-slice, and fluoroscopic CT), and clinical practice. Studies in the UK suggest as an initial trend broadly increasing levels of exposure per examination; the overall mean doses per CT examination from regional surveys in Wales, 1994, and Northern Ireland, 1996, were 20% and 5% higher, respectively, than the level observed in a national survey for the UK in 1989 (Clarke et al., 2000).

(31) On the basis of equivalent scanning parameters, doses from spiral scanning are broadly similar to those from serial scanning, although increases by 10–30% will occur with multi-slice detector-array scanners (Kalender, 2000). Such technology can provide reductions in dose by the use of an increased pitch (>1), yet could also stimulate increased complexity of the examination and overall increases in patient dose. CT fluoroscopy is conducted at lower tube currents than for conventional scanning, although patients may remain stationary in the x-ray beam for significant periods of time. Absorbed dose rates to the skin are typically 2- 8 mGy per second, with effective dose rates of 0.03–0.07 mSv per second when scanning at the level of the mid-abdomen (Keat, 2000). Typically, conventional CT imparts a localised absorbed dose of 20–50 mGy to the surface of the body.

(32) Use of electron beam CT (EBCT) has been primarily limited to cardiac application, however newer machines have increased the number of images available and applications outside the heart are feasible. Using 3 mm collimation, $CTDI_w$ is virtually identical (5.0 mGy) for EBCT and spiral CT, whereas for 1.5 and 6 mm collimation, EBCT has a 75% and 106% higher average dose in comparison to the 1 and 7 mm spiral CT collimation (Weisser, 1999).

(33) Notwithstanding the mean levels of dose in CT discussed above, surveys of clinical practice have also demonstrated wide variations in patient dose and potential for improvements in optimisation of the examination. Typical doses for a given general type of procedure have been shown to vary between individual CT centres by factors of 10–40 in the UK (Shrimpton et al., 1991) and 8–20 in Norway (Olerud, 1997). Such variations were largely due to differences in the local scanning technique employed, such as the number and thickness of slices imaged in serial scanning, the use of contrast medium for additional scans, and the exposure settings selected. There is thus a continuing need for critical review of current practice, more widespread assessment of dose, and the use of reference doses.

3. WHAT PRACTICAL ACTIONS CAN BE USED TO MANAGE PATIENT DOSE?

(34) The referring physician should evaluate whether the result of each examination will affect patient management. The radiologist should be satisfied that the procedure is justified. The operator should be aware of the possibilities to reduce patient doses by adapting technical parameters to each patient and the examination at hand, with special attention being paid to paediatric and young patients. More than a 50 percent reduction in patient dose is possible by an appropriate choice of technical parameters, attention to quality control, and the application of diagnostic reference levels in co-operation with a medical physicist.

3.1. Introduction

(35) Radiologists and referring clinicians have a critical role in ensuring that patients are not irradiated unjustifiably. This section reviews the steps that referring clinicians and radiologists should undertake to discharge their responsibilities satisfactorily. It should be noted that in some countries this concept may be embodied in national law. The observations made in this section assume that medical practitioners will be fully familiar with regulatory and advisory requirements in their country.

3.2. Justification

(36) Requests for a CT examination should be generated only by properly qualified medical or dental practitioners depending on the national educational and qualification system. The radiologist should be appropriately trained and skilled in computed tomography and radiation protection, and possess adequate knowledge concerning alternative techniques. A fundamental principle of radiation protection is that of justification, under which no investigation is undertaken unless the radiation dose is deemed to be justified by the potential clinical benefit to the patient. Also to be considered in the justification process are the availability of resources and cost. Justification is a shared responsibility between clinician and radiologist.

(37) Clinical guidelines advising which examinations are appropriate and acceptable should be available to clinicians and radiologists. Ideally, these will be agreed at the national level, but where they are not, local guidelines are often developed within an institution. Where possible, clinically relevant examinations should be obtained with the lowest achievable radiation dose to the patient consistent with obtaining the diagnostic information. In CT, this requires consideration of whether the required information could be obtained by conventional radiography, ultrasound, or magnetic resonance imaging (MRI) without unduly hindering clinical management.

(38) Where CT is deemed to be justifiable clinically, consideration must be given to tailoring the examination to diagnostic needs of the patient. This is good practice

and constitutes one of the most important protection roles of the radiologist. CT scanning in pregnancy often raises concern. CT scanning of pregnant females may not be contra-indicated, particularly in emergency situations, although examinations of the abdomen or pelvis should be carefully justified. For computed tomography scans with the uterus in the field of view, the absorbed doses to the fetus are typically about 40 mGy. Fortunately, the primary radiation beam on CT scanners is very tightly collimated and can be precisely controlled relative to location by using scout view (topogram). As with other examinations it may be possible to limit the scanning to the anatomical area of interest (ICRP, 2000).

(39) As in all x-ray procedures, CT examinations should not be repeated without clinical justification and should be limited to the area of pathology under request. Unjustifiable repetition of exposure may occur if the referring clinician or radiologist is unaware of the existence or results of previous examinations. The risk of repetitive examinations increases when patients are transferred between institutions. For this reason, a record of previous investigation should be available to all those generating or carrying out examination requests. The clinician who has knowledge that a previous examination exists has a responsibility to communicate this to the radiologist.

(40) CT examinations for research purposes that do not have clinical justification at the level of immediate benefit to the person undergoing the examination should be subject to critical evaluation since the doses are significantly higher than conventional radiography. Additional information on this is available in *Publication 62* (ICRP, 1993).

3.3. Managing the patient dose

3.3.1. Optimisation

(41) Once referral for CT examination has been justified, the radiologist has primary responsibility for ensuring that the examination is carried out conscientiously, effectively, and with good technique. This is usually described as the principle of optimisation. Within this process the radiologist and operator have considerable scope for limiting the radiation dose to the patient. The objective is to provide sufficient diagnostic information to influence the clinical management of the patient. Clinical issues define the area to be examined and the extent of the examination required. However, even when these conditions are met, the radiologist has additional opportunity for limiting the radiation dose to the patient.

(42) It is valuable to consider the role of contrast medium enhancement prior to commencing the examination. In some cases, a single examination following enhancement may be adequate for clinical purposes and initial unenhanced images may therefore be avoided. In multiphase enhancement studies, the examination should be limited to the number of phases which are clinically justified.

(43) CT fluoroscopy and interventional CT pose particular challenges in radiation protection. Conventionally, biopsy procedures have often been performed with x-ray fluoroscopic or ultrasonographic guidance. However x-ray fluoroscopy supplies

limited 3-dimensional information and ultrasound guidance may be impeded by bowel gas, lung, or bone. For this reason, CT-guided percutaneous biopsy is widely performed and has the advantage of operational ease and safety. However this involves a longer exposure and the patient and radiologist may be exposed to high doses of radiation.

(44) A number of national surveys have indicated widespread variation in the radiation dose to patients for any particular radiological examination (Shrimpton et al., 1991; Conway et al., 1992, Hart et al., 1996). In conventional radiography, higher exposure leads to increased darkening of the image, whereas in CT that is not the case and this can result in the selection of unnecessarily high exposure factors (Rehani, 2000; Rehani and Berry, 2000). There is much that the radiologist and the technologist can do to keep radiation exposures low without compromising image quality. The operator has control over tube current (mA), scan length, slice thickness (collimation), table feed per 360° (pitch), and applied potential (kVp). Commonly CT machines provide pre-set factors; however, the settings should be tailored for each patient according to body part and patient build. Protocols should be designed to include patient parameters and established in co-operation with a medical physicist.

3.3.2. Role of mA and mAs

(45) The mAs is the single most important factor for managing patient dose. It should vary with patient size and body part. Reducing mAs significantly reduces patient dose and lengthens tube life. The mA controls the x-ray intensity (the number of x-ray photons per unit time). The mAs setting represents the number of x-ray photons in the defined exposure time. The intensity is directly proportional to mA. For a given scanner, halving mA means halving radiation dose rate and halving mAs means halving radiation dose. Since its invention about 3 decades ago, the trend in all subsequent developments in CT has been to minimise scanning time. When the image of a defined region is to be acquired in seconds or even fractions of a second, high x-ray intensity is a must. The shorter the exposure time, the higher the required x-ray intensity. Accordingly, x-ray tubes for CT are designed to give better radiation output and improved heat capacity and heat dissipation.

(46) Another factor that contributes to high dose in CT is the demand for higher spatial resolution, leading to the use of thinner sections, which in turn necessitates even higher intensities of x-ray beam in order to keep noise low. High resolution CT requires thin slices typically of 1 or 2 mm, which is only possible by increasing the mA. For a fixed value of mAs, decreasing exposure time (s) means proportionately increasing the tube current (mA). Reduction of mA without corresponding increase in exposure time leads to an increase in noise and thus a possible degradation in image quality. In good practice, one should strike the balance between image quality and dose.

(47) The degradation in image quality as a result of reducing mAs is not significant in high contrast situations. In the body there are some high contrast structures, like the thorax and pelvis, where the contrast between bony structures and

soft tissue or air is high. In such situations, a significant decrease in mA is possible while keeping image quality acceptable. This strategy has been exploited by many investigators, particularly in relation to imaging of the thorax. For example, a low dose CT technique of the thorax was described in 1990 whereby scans of acceptable diagnostic quality were obtained with an mAs setting that was only 20% of that used for standard practice (Naidich et al., 1990). For chronic infiltrative lung disease, a high confidence level in diagnosis has been demonstrated in 61% of low dose CT scans compared with 63% of conventional dose CT scans (Lee et al., 1994). Studies under simulated conditions using phantoms demonstrated that there is no decrease in detection of simulated plaques, nodes, and effusions in a chest phantom when mA is reduced by 80%, typically from 400 to 80 mA (Mayo et al., 1995).

(48) A similar reduction in absorbed dose in paediatric chest CT has been reported. A low dose technique using 25 mAs (in a typical case) has been shown to provide image quality with no loss of diagnostic information (Rogalla et al.1999). In an attempt to find minimum tube current for spiral CT, the subjective quality of images obtained at 20 mAs has been reported to be not significantly different from that assessed for images obtained at 50 mAs. Imaging of the middle zone of the chest requires still lower values of mAs (approx. 12 mAs), relative to the upper and lower zone, which require around 20 mAs. (Itoh et al., 2000). It is possible to perform spiral CT of the maxilla and mandible with a radiation dose similar to that used for conventional panoramic radiography.

(49) There are definite problems in achieving low doses in areas of low contrast in the body, like the abdomen. Noise becomes a limiting factor in such circumstances. It is a common practice to use the same mAs whenever abdomen and pelvis are to be scanned. Substantial dose reduction, without any recognisable deterioration in diagnostic image quality, may be achieved if pelvic CT is performed at almost $1/3^{rd}$ the mAs for abdomen region. A surface dose reduction from 30 to 10 mGy has been documented. The rationale behind reducing the mAs for imaging of the pelvis relative to the abdomen is that the abdomen contains organs like the liver, where resolution is very important, whereas the pelvis does not have similar structures, but rather bones, bladder, and opacified bowel. An increase in mAs does not significantly improve high contrast resolution, but leads to a major change in low-contrast areas. Thus lower mAs values may not create problems for imaging of the pelvis, but are not desirable for the abdomen.

3.3.3. mA modulation

(50) Recently, attempts have been made to develop the so-called '*smart technique*' with the principal idea being to change technical factors during a 360° rotation and along the Z-axis according to the actual object attenuation, instead of keeping tube current constant for all projection angles and Z-axis positions as is usual practice today (Kalender, Wolf, Suess et al., 1999). If the manufacturers implement this, it will contribute in a large measure to reduction in patient dose and the need for subjective adjustment of mA may be reduced. Further details are given in Chapter 4.

3.3.4. Scan length

(51) This controls the volume of patient irradiated. Unfortunately, with the advent of fast scanners, there is a tendency to increase the scan length so much that examinations of the thorax + abdomen + pelvis are becoming much more common. Practice may soon include head-to-pelvis examinations (particularly for rapid assessment of patients with massive trauma). It is essential to draw the attention of referring physicians and radiologists to the dose consequences of such practices and efforts must be made to restrict the areas of examination to those clinically essential.

3.3.5. Collimation, table speed, and pitch

(52) In conventional CT, the table speed and pitch are absent. In spiral CT, however, all three factors have to be considered together. They are inter-linked in such a way that discussion of one in isolation is not sufficient to characterise scanning movement and patient dose. For example, pitch is table feed (mm) in one rotation relative to collimation (slice thickness). If the pitch is taken as 1, it can be achieved by 10 mm/rotation for 10-mm collimation. If the rotation time is one second for 360°, the table speed (table feed per rotation) becomes 10 mm/sec. If one alters the collimation to 5 mm without changing table speed, the pitch becomes 2. Pitch changes have different effects on image quality in different situations. For some situations, like in virtual CT coloscopy, image quality and reconstruction artefacts are less affected by the pitch value than by beam collimation. Thus from an image quality point of view, one may prefer a higher pitch with narrow beam collimation. But the situation is different for small pulmonary nodules, which may require thin section CT (lower collimation) and where an increased pitch may affect detectability. Keeping a pitch of 1 while using thinner sections results in higher radiation dose.

(53) There are two ways by which pitch can be increased for a constant tube rotation: increase table travel speed or decrease collimation. These methods have different effects:

(a) Increasing table travel speed for a given collimation and hence higher pitch is associated with lower radiation dose over the scanned volume for constant mA (due to lower effective exposure time) and predictably decreased detection of lesions like small pulmonary nodules.
(b) Decreasing collimation (for a given table speed) results in unchanged scan time, decreased radiation dose over the scanned volume for constant mA, and decreased signal-to-noise ratio; however, the detection of small pulmonary nodules is potentially superior due to mA and the signal-to-noise ratio.

(54) Contiguous spiral CT scans (pitch of 1) give approximately the same radiation dose as contiguous axial scans acquired with the same technical factors. For non-contiguous scans (pitch > 1) at a given collimation, the radiation dose decreases as the pitch increases, specifically as 1/pitch. For a given table speed of 10 mm/s, the

radiation dose from a 10 mm collimation scan (pitch of 1) is approximately double that of a 5 mm collimation scan (pitch of 2). For a given collimation, increasing the table speed (increasing the pitch) reduces the radiation dose by 1/pitch. For example, going from 10 mm and pitch of 1 (10 mm/s) to 10 mm and pitch of 2 (20 mm/s) reduces the radiation dose by 50%. At very small slice thickness, the radiation profile width (full width at half maximum) may be greater than the nominal slice thickness, which results in extended radiation overlap between contiguous slices (and no net change in radiation dose compared with thicker slices). While thinner collimation would normally be expected to yield a smaller radiation dose, the higher degree of overlap between adjacent scans offsets this expected decrease and ultimately results in little net effect on absorbed dose due to collimation.

3.3.6. Role of combination of factors

(55) Normally, kVp is not changed from patient to patient for a particular type of study, even though many machines make it possible to change the setting and it may be desirable to do so. Assuming that scan length and slice thickness have been judiciously chosen as per clinical need, one is left with mA, table feed/rotation and pitch. Table 4 gives a typical example of settings for spiral CT of the chest, in which mA has been reduced from 165 to 110, table feed increased from 5 mm to 10 mm per rotation, and pitch changed from 1 to 2 (Kalender, Schmitt, Zankl et al., 1999). This results in the effective dose decreasing from 7.1 to 2.4 mSv (i.e. 34% of original or a 66% reduction) and lung dose decreasing from 24.3 to 8.2 mGy (66% reduction). A similar example for quantitative CT of the lumbar spine indicates a 92% reduction in absorbed dose by reducing kVp and mA (Table 4).

(56) Facial CT is used for osseointegrated implants and can be performed with spiral CT and a dental software package. Reducing mAs from 165 to 35 and using a pitch of 2 rather than 1 has been reported to reduce the eye lens dose by a factor of nearly 2 (e.g. 0.5 mGy to 0.3 mGy), thyroid gland dose by a factor of 5 (e.g. 2.5 mGy

Table 4. Examples for dose reduction in CT by changes in scan parameters from set 'a' to set 'b' (adapted from Kalender, Schmitt, Zankl et al., 1999)

Parameter	Spiral CT of the chest [a]	Spiral CT of the chest [b]	Quantitative CT of the lumbar spine [a]	Quantitative CT of the lumbar spine [b]
Voltage (kVp)	140.0	140.0	120.0	80.0
Current (mA)	165.0	110.0	250.0	75.0
Scan range (cm)	31.0	31.0	3 × 1.0	3 × 1.0
Slice thickness (mm)	5.0	5.0	10.0	10.0
Table feed/360° (mm)	5.0	10.0	–	–
Pitch	1.0	2.0	–	–
Organ of interest	Lung		Stomach	
Organ dose (mGy)	24.3	8.15	4.4	0.4
Effective dose (mSv)	7.1	2.4	1.0	0.1

to 0.5 mGy), and parotid gland dose by a factor of 6 (e.g. 2.4 mGy to 0.4 mGy). These reductions in dose did not lead to any significant loss of image quality or diagnostic information (Rustmeyer et al., 1999).

3.3.7. Shielding of superficial organs

(57) Conventionally, organ shielding has not been practised in CT. However, increased doses in CT have generated interest in this area. Shielding is particularly relevant in children. Use of shielding should not be an excuse to raise exposure parameters. Breast, thyroid, lens of the eye, and gonads are seldom the organ of interest in a CT examination, although they are often incidentally in the beam. The radiation doses delivered to these organs are significant enough to be a matter of concern.

(58) A conventional diagnostic chest CT imparts a dose of 20-50 mGy to the breast of an averaged sized woman. This is equivalent to 10-25 two-view mammographic examinations. CT examinations of the chest in girls and young females need to be justified in view of the higher risks of radiogenic breast cancer for this age group. Shielding of breast tissue by a breast garment of thinly layered bismuth impregnated radioprotective latex has been shown to reduce the radiation dose by over 50% without affecting the display of other deeper structures (Hopper et al., 1997). Whether to use bismuth or lead is to be decided on the basis of ease of manufacturing, versatility, fit, and cost.

(59) CT slices at the base of the skull impart high doses to the thyroid and shielding of this organ in children is very effective in such cases. The dose to the eye lens is typically around 30 mGy in general head CT, 70 mGy in scanning of sinuses, and may be 10-130 mGy in CT of orbital trauma. Proper angling of the gantry helps in dose reduction to the eye lens. Further, the use of CT fluoroscopy should be limited to the areas where the technique has the potential to provide useful information only.

(60) Gonadal shielding during CT examinations is controversial. When the gonads are not included in the examination field, the small doses are due to internal scatter and thus external shielding is largely ineffective. When the gonads are within the direct CT beam, shielding may be considered if they are not the organ of clinical concern and if shielding will not compromise the examination, by producing significant artefacts or by directly obscuring a contiguous area of clinical interest. Shielding of the ovaries is difficult because their exact location is usually not clear and the expected pathology is often nearby.

3.3.8. Partial rotation

(61) In CT, the x-ray tube rotates around the patient resulting in cross-sectional images. The speed of rotation and the scan parameters like kVp and mAs are constant throughout the 360° rotational path in the models currently available commercially. The dose at the surface of the patient thus depends upon the distance from the x-ray target for the entrance beam, together with contributions from the

primary and secondary x-ray photons from the beams entering the patient from other points. The major contribution to the absorbed dose at any point superficially located arises from the entrance dose. For the eye lens, the frontal beams thus contribute the major part of the absorbed dose. In head scanning, if the frontal 90° is omitted and the scan is performed with an angular rotation of 270°, then as shown in Fig. 2 the eyes receive a minimal dose (Robinson, 1997). This partial rotation capability is currently available in some scanners.

3.4. Dose in CT fluoroscopy

(62) Unlike x-ray fluoroscopic or ultrasound (US) guidance, in which the biopsy needle and the lesion can be observed in real time, guidance with conventional CT does not permit imaging during the actual procedure, which remains 'blind'. This limitation of conventional CT-guided procedures has been overcome by the development of continuous imaging and CT fluoroscopy which permit CT images to be reconstructed and displayed with a reconstruction time of less than 0.2 s. CT fluoroscopy allows tomographic images to be observed in real time as an animated sequence.

(63) During conventional CT-guided biopsy, the physician is not exposed to x rays during scanning. One major problem with CT fluoroscopy-guided procedures, however, is that the physician's hands are exposed to high levels of radiation because various procedures are performed within the direct x-ray beam when the interventional devices are manipulated manually. Direct beam exposure to the hands may reach 120 mGy per procedure if protection is not provided. This would limit the number of procedures that a physician could perform to 4 per year (given the dose limit of 500 mSv to hands recommended by the Commission; ICRP, 1991).

(64) Since the beam in CT is finely collimated, effective protection can be achieved by moving the hands slightly so as to be out of the collimated primary beam. This can be achieved with the use of holding instruments for the syringe, needle, etc. The holder should be of acrylic so as to prevent the streak artefacts associated with a metal holder. The hands can thus be 5 cm or so away from the primary beam and are thus exposed to scattered x rays only. The exposure is reduced by over 98%. The success rate reported with the use of a holder is 100% and no significant increase in operating time has been observed (Kato et al., 1996). The absorbed doses to patients are high as indicated in Chapter 2. The most effective way to control the absorbed dose is to minimise the fluoroscopy time and attention must be focused on this parameter.

3.5. European Commission quality criteria

(65) Recommendations concerning achievable standards of good practice in CT have been developed by the European Commission in the form of quality criteria (CEC, 1999). This concept seeks to provide an operational framework for radiological protection initiatives in which technical parameters for image quality are considered in relation to patient dose. Diagnostic and dose requirements for CT are

specified in terms of the quality criteria considered necessary to produce images of standard quality for a particular anatomical region, without regard to specific clinical indications.

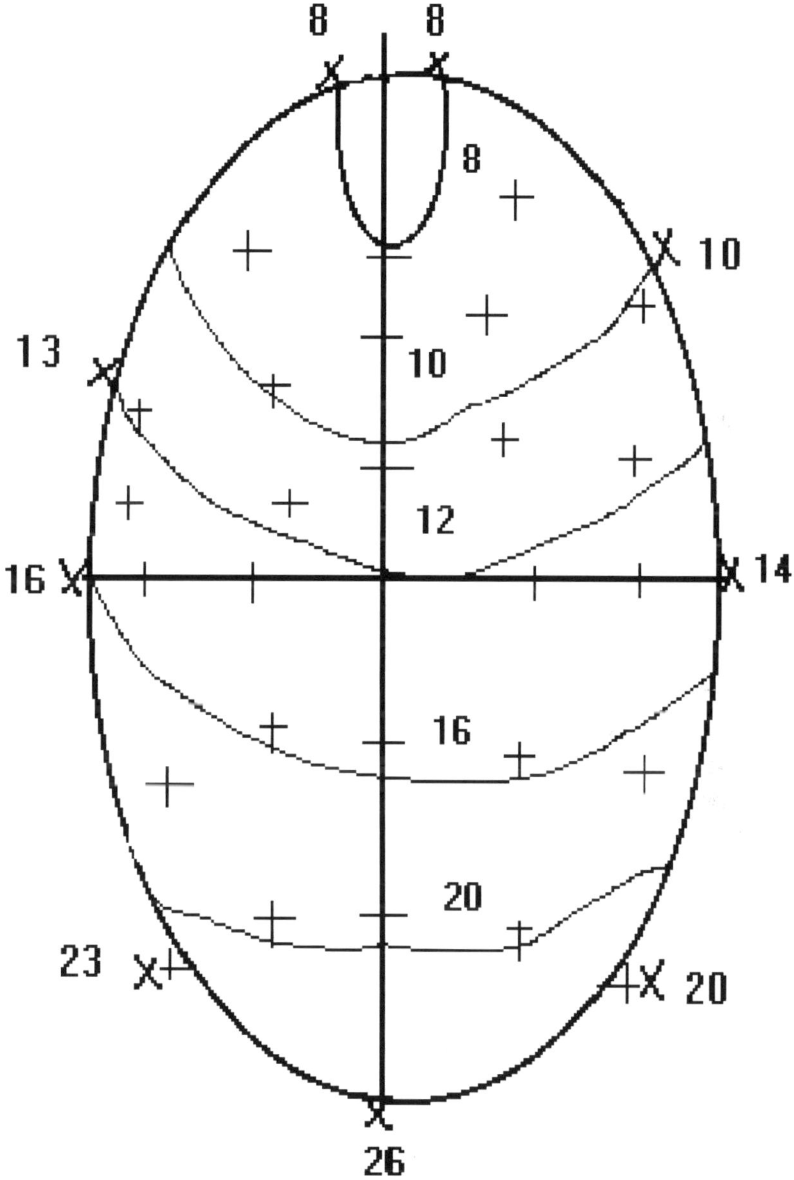

Fig. 2. Dose distribution (in relative units) through the section of the skull (face-up) for a 270° scan omitting the frontal 90°. Minimum dose occurs in the region of the eyes. The doses are slightly higher on the left side since in this unit, the x-ray tube rotates by an additional 20 °C (clockwise) for patient movement. Adapted from Robinson (1997).

(66) The subjective image criteria include anatomical criteria that relate to the visualisation or critical reproduction of anatomical features. Criteria concerning patient dose are given in terms of reference dose values associated with the examination technique used for standard-sized patients. Quality criteria have been developed for 26 types of examinations within 6 broad anatomical groups, together with examples of technique parameters influencing the dose. The usefulness of this framework for detailed audits of CT practice has been investigated and demonstrated in clinical trials (Calzado et al., 2000; Jurik et al., 1998).

3.6. Dosimetry

3.6.1. Diagnostic reference levels

(67) The diagnostic reference level is an essential element of quality assurance in CT. The implementation of DRLs is done through measurement of CTDI, performed in standard dosimetry phantoms, using a pencil ionisation chamber with an active length of 100 mm and calibrated in terms of absorbed dose to air (see Chapter 2 and Appendix A). These measurements should be conducted as part of routine constancy testing (quality control) for each scanner (IPEM, 1997; IEC, 1999). To avoid a potential source of confusion, it should be recognised that previous recommendations published in the literature have sometimes utilised subtly different definitions of CTDI involving, for example, different lengths of integration or reference materials. (IEC, 1994; Edyvean, 1998).

(68) Measurement of CTDI in the standard head and body dosimetry phantoms allows the derivation of the quantities $CTDI_w$ and DLP for any given clinical scanning protocol (CEC, 1999). In the absence of measurements for an individual scanner, broad estimates of dose may be made using model-specific generic data from published compilations of CTDI data (CEC, 1999). Assessments of typical CT practice should be based on local surveys involving the evaluation of scan details for representative samples of at least 10 patients for each type of procedure. Mean values of $CTDI_w$ and DLP for each patient group should be compared with appropriate diagnostic reference levels that have been set nationally or locally to promote optimisation of patient protection (Appendix A). Any technique for which doses (DRL) are above an investigation level should be critically reviewed and either clinically justified or revised so as to reduce patient doses without loss of clinical efficacy. Such assessments should be carried out periodically (for example, at least every 3 years) or whenever there are substantial changes to equipment or technique.

3.6.2. Effective dose

(69) There is also a need for CT facilities to know typical effective doses for the different common types of procedure in clinical use. Such doses can be estimated from values of DLP or calculated with knowledge of the scanning technique, using published dose coefficients and CTDI measurements made free-in-air (Chapter 2).

4. WHAT NEW EQUIPMENT FEATURES WOULD HELP MANAGE PATIENT DOSE?

(70) CT doses are relatively high and have not decreased over time as they have in conventional radiography. Further improvements in CT equipment could help the operator to reduce unnecessary patient dose substantially. The most important of these features will be anatomically based on-line adjustment of exposure factors and new image reconstruction approaches associated with multi-slice computed tomography.

4.1. Introduction

(71) The frequency of CT examinations has continued to increase in recent years, in spite of the widespread availability of MR imaging. In view of the newest technical developments in CT, it is not expected that this trend will reverse in the foreseeable future. On the contrary, the very possibilities which multi-slice spiral CT offers will most likely lead to a further increase in the number of CT examinations performed. Consequently, it also appears likely that the collective dose to the general population from CT will remain at the present level or increase accordingly. The relative contribution of CT to the annual exposure of the population to ionising radiation for medical diagnosis will continue to increase. A meaningful assessment of CT can take place only if, along with the discussion of possible risks, the clinical benefits, i.e., diagnostic reliability, patient comfort, costs, etc. are also taken into consideration. Such a detailed discussion would go beyond the framework of this document.

(72) During nearly 30 years of clinical use of CT, significant improvements have been observed in image quality and general scanning performance. Among the many parameters of performance, reductions in the scan times per slice, i.e. the rotation time per 360°, and the scan times per volume of examination have been the most impressive changes. CT now routinely offers sub-second rotation times and total examination times of 10 to 60 s. In comparison, the improvements in low-contrast resolution have been less spectacular. The dose per slice for given levels of noise and resolution has not changed dramatically over the past years.

(73) CT developments in the past always aimed at improving utility, i.e. at enhancing the diagnostic value of established CT applications and at providing new applications. Dose efficiency was not a primary goal to the same degree, and the 'market' did not demand that manufacturers pursue it. Consequently, there still appear to be a number of possibilities to optimise CT systems and their use with respect to dose efficiency. It has to be stressed that efforts have to be supported by both the manufacturers and the users of the systems, respectively. Some important points are summarised in Table 5.

4.2. Spiral CT

(74) Spiral CT offers specific possibilities for the reduction of dose. A very effective method for dose reduction is given by choosing a pitch factor of greater than 1. The

Table 5
Possibilities for patient dose reduction with CT

Measures for the user	[Measures for the manufacturer]
Checking the indication and limiting the scanned volume	Increasing the pre-filtration of the radiation spectrum
Adapting the scanning parameters to the patient cross-section	Attenuation-dependent tube current modulation
Pronounced reduction of mAs values for children	Low-dose scanning protocols for children and special indications
Use of spiral CT with pitch factors >1 and calculation of overlapping images instead of acquiring overlapping single scans	Automatic exposure control for conventional CT and spiral CT
Adequate selection of image reconstruction parameters	Noise-reducing image reconstruction procedures
Use of z-filtering with multi-slice CT systems	Further development of algorithms for z-filtering and adaptive filtering

specific use of the new possibilities which multi-slice CT systems offer can also serve to limit the dose. The new approaches to z-interpolation and z-filtering, which allow for retrospective variation of the effective slice thickness, provide images both with high 3D spatial resolution or alternatively with low noise and excellent low-contrast resolution without the need for additional exposure to radiation.

4.3. Tube current, filtration, and other technical factors

(75) Some technical measures to improve the dose efficiency of CT systems are known and have been partly tested. However, their use often creates a conflict in relation to other goals and requirements. Thus, for example, increasing the filtration, which reduces the patient dose, requires higher mAs values and thus leads to greater loading of the x-ray tube. This in turn can lead to limitations of the permissible scan duration for spiral CT. Multi-slice CT systems drastically reduce the scan duration, and the decreased loading of the tube can permit the use of additional filtration.

(76) The definition and preparation of low-dose scanning protocols for paediatric CT and for special indications should be studied further and be promoted actively by the manufacturers. The further development of noise-reducing reconstruction methods also appears to be promising, especially with regard to multidimensional adaptive filters for multi-slice CT systems, which offer considerable potential (Kalender, 2000; Kachelriess, 2001).

(77) A significant reduction in dose can be achieved through anatomy-adapted, attenuation-dependent modulation of tube current. The basic idea is that the pixel noise in a CT image is largely attributable to those projections in which the attenuation and therefore the quantum noise are greatest. This means that for cross-sections deviating significantly from a round shape, the intensity of the radiation can be reduced in the projections with less attenuation without any significant effect on

the noise pattern. This offers considerable potential for the reduction of dose without a deterioration of image quality, as has been shown clearly in several studies (Gies et al., 1999; Kalender, Wolf and Suess, 1999; Kalender, Wolff, Suess et al., 1999.

(78) Human anatomy practically always involves cross-sections that deviate more or less significantly from a circular or cylindrical shape. Accordingly, studies with tube current modulation show that the mAs product can be reduced typically by between 10% and 50% without any loss of image quality. For scans with extreme differences in the attenuation characteristics between the antero-posterior (AP) and lateral directions, such as in the shoulder region, absorbed dose reduction of more than 50% is possible (Greess et al., 1999; Kalender, Wolf and Suess, 1999).

(79) With the use of tube current modulation, it is also possible to influence selectively image quality. Increasing the tube current in the lateral direction and reducing the current in the AP direction can improve the image quality and at the same time significantly reduce the dose. The actual patient dose is reduced even more than the mAs product. In hip examinations, for example, typical values of mAs reduction of around 40% were found. These corresponded to a reduction in patient dose of nearly 60 to 70% as determined by phantom measurements and by Monte Carlo calculations (Kalender, 2000).

(80) Spiral CT is predestined for the novel tube current modulation techniques, since the required reference data on attenuation for modifying the tube current are available in the shortest possible times and over the shortest possible distances. The currently used modulation parameters are determined in real-time from the immediately preceding values, taken from the oppositely located tube position, i.e. shifted by only 180° or half the table feed per rotation. This approach is generally applicable in spiral scanning. All manufacturers should be encouraged to implement and to offer it. A beneficial side effect is that tube life should go up or that the demands on the x-ray components may be reduced as total mAs per examination goes down.

4.4. Providing information on dose

(81) Increased efforts in performing routine dosimetry at CT facilities and establishing reference dose levels will help not only to control and optimise techniques, but also to make information on the orders of magnitude of patient dose more generally known.

(82) Indication of CTDI values and DLP information by the manufacturers on the operator's console is a valuable step in that direction, and the radiologist and operator should be familiar with the presentation and the meaning of these doses. There are also established procedures and software to provide estimates also of organ and effective doses for a typical patient, i.e. for 'standard man' (Zankl et al., 1991; Jones and Shrimpton, 1993; Kalender, Schmitt, Zankl et al., 1999). It is also possible to calculate the dose distribution specific to the patient and scan protocol by Monte Carlo methods (Kalender, 2000).

4.5. Automatic exposure control (AEC) for computed tomography

(83) Optimisation of CT systems and quality control have to ensure that a diagnostic quality image is obtained with a minimum of dose. For this to occur, existing measures must be supplemented by two further steps, which will require the close cooperation of manufacturers and users: development of an automatic exposure control for CT and objectively defined requirements for image quality. The combination of these two measures seeks to achieve and secure as its goal a definite level of image quality, attainable with minimum dose for the particular examination type. This would also include definition of standards and diagnostic reference levels.

(84) The development of anatomy-dependent, attenuation-based methods of x-ray tube current regulation have shown a high potential for dose reduction (Greess et al., 1999, Kalender, Wolf and Suess, 1999; Kalender, Wolf, Suess et al., 1999). These methods should be used on a broad basis, since they do not entail any disadvantages in terms of image quality. Their introduction on a general basis, however, should only represent a first step, because they are so far limited to the optimum distribution of a predefined mAs product per 360°. A further necessary step must be the development of an automatic exposure control for CT. Automatic exposure controls have been established since long in conventional x-ray diagnostics, although less for the purpose of dose reduction than for the prevention of faulty exposures. In CT, the idea of an automatic exposure control has not yet been followed up because, as a result of the prevailing high dynamic ranges of the receptor systems, faulty exposures in the classical sense can be ruled out and because certain technical questions remain to be clarified.

(85) In the meantime, technical possibilities, such as anatomy-dependent tube current regulation, are available. The remaining technical problem for the implementation of an AEC lies in the fact that, along with the regulation of the tube current during a 360° rotation, the tube current- time product (that is, the mAs value per rotation) during a spiral CT examination continuously adapts to the changing body cross-section and the particular attenuation. This is technically feasible, provided that certain limiting conditions are considered, such as limitation of the maximum tube current. The essential problem is to arrive at an objective pre-setting to define the image quality, defined for example in terms of image noise and image sharpness, required for a specific examination type. The problem of calculating the required tube current values from the measured CT data in real time can be solved. The requirements for image quality must, however, first be defined by radiologists.

4.6. Image quality

(86) Objective measures of image quality are available. Nevertheless, experience shows that these cannot lay claim to a complete description of images. This refers, for example, to noise patterns, which can be influenced by both the dose and the choice of convolution kernel. The subjective assessment of image quality by the radiologist can very easily differ from an objectively determined order of ranking. In

spite of this, it should be possible to arrive at a consensus with respect to the decisive parameters, above all for image sharpness and noise.

(87) This image quality which is to be seen as the 'minimum necessary' or, in the sense of radiation protection, as the 'optimum' image quality for a particular application, must be ensured for all patients and without exception for all slices of the volume to be examined, with minimum dose. The dose would then automatically be reduced for the examination of a slender patient. This of course applies in special measure to paediatric CT, where minimum dose with the assurance of acceptable image quality is a matter of greatest importance. The dose would likewise be automatically reduced when, in the course of an examination, thinner cross-sections are reached; something that is almost never realised under the conditions prevailing today.

(88) A consensus concerning the required image quality parameters for commonly performed CT examinations would also improve the situation of different institutions working with substantially different parameters and consequently delivering different absorbed doses. Differences of more than a factor of four have frequently been reported (Shrimpton et al., 1998). This would also simplify the comparison of different CT scanners with regard to their dose requirements and enable the definition of acceptance criteria for general use.

4.7. Potential for accidents

(89) There have been almost no accidents involving CT scanners that have resulted in high absorbed doses. This is due to scanner design. The most obvious opportunity for accidental exposure in spiral CT would be if the table were mechanically jammed, did not move, and the tube continued rotational exposure. Mechanical jamming of tables is actually reasonably common with patient restraint devices, sheets, and tubes getting caught under the table. Fortunately, the scanners are equipped with linear resistive potentiometers that sense the rate of change of table velocity, and if there is a discrepancy of table velocity from that expected for the set pitch etc., the exposure is immediately terminated. In the design of new equipment, manufacturers need to continue to assess prospectively the potential for accidents.

Annals of the ICRP

APPENDIX A. REFERENCE DOSE QUANTITIES FOR COMPUTED TOMOGRAPHY

(A 1) The principal dosimetric quantity used in CT is the computed tomography dose index (CTDI). This is defined as the integral along a line parallel to the axis of rotation (z) of the dose profile (D(z)) for a single rotation and a fixed table position, divided by the nominal thickness of the x-ray beam. CTDI can be conveniently assessed using a pencil ionisation chamber with an active length of 100 mm, so as to provide a measurement of $CTDI_{100}$, expressed in terms of absorbed dose to air (IEC, 1999):

$$CTDI_{100} = \frac{1}{nT}\int_{-50}^{+50} D(z)dz \qquad \text{(mGy)} \qquad (1)$$

where n is the number of tomographic sections, each of nominal thickness T, from a single rotation.

(A 2) Reference dosimetry for CT is based on such measurements made within standard CT dosimetry phantoms. These presently comprise homogeneous cylinders of polymethylmethacrylate (PMMA), with diameters of 16 cm (head) and 32 cm (body), although phantoms of water-equivalent plastic and with elliptical cross-sections are under development. The combination of measurements made at the centre (c) and 10 mm below the surface (p) of a phantom leads to the following two reference dose quantities (CEC, 1999):

(a) Weighted CTDI in the standard head or body phantom for a single rotation corresponding to the exposure settings used in clinical practice:

$$CTDI_w = \frac{1}{3}CTDI_{100,c} + \frac{2}{3}CTDI_{100,p} \qquad \text{(mGy)} \qquad (2)$$

where $CTDI_{100,p}$ represents an average of measurements at four different locations around the periphery of the phantom.

(b) Dose-length product for a complete examination:

$$DLP = \sum_i {}_nCTDI_W * nT * N * C \qquad \text{(mGy cm)} \qquad (3)$$

where i is the number of scan sequences in the examination, each with N rotations of nominal beam collimation nT cm [as defined in Equation (1) above] and exposure C

mAs per rotation; $_{n}CTDI_{w}$ is the normalised weighted CTDI (mGy $mA^{-1}s^{-1}$) appropriate for the applied potential and total nominal beam collimation.

(A3) These quantities can be applied to serial or spiral scanning, for both single- or multi-slice geometry scanners. Initial diagnostic reference levels have been published for some common procedures on the basis of surveys of practice for adult (CEC, 1999) and paediatric (Shrimpton and Wall, 2000) patients; these values are shown in Tables A1 and A2. Such investigation levels are for comparison locally with the mean values of dose descriptors assessed in a CT facility during examinations on representative groups of patients and should not be applied on an individual patient basis.

Table A1. Initial diagnostic reference levels for CT examinations on adult patients (CEC, 1999)

Examination	Diagnostic reference level*	
	$CTDI_w$ per slice or rotation (mGy)	DLP per examination (mGy cm)
Routine head	60	1050
Face and sinuses	35	360
Vertebral trauma	70	460
Routine chest	30	650
HRCT of lung	35	280
Routine abdomen	35	780
Liver and spleen	35	900
Routine pelvis	35	570
Osseous pelvis	25	520

*Same as ICRP reference levels.

Table A2. Initial reference dose values for CT examinations on paediatric patients (Shrimpton and Wall, 2000)

Examination	Patient age (years)	$CTDI_w$ per slice or rotation (mGy)	DLP per examination (mGy cm)
Brain	<1	40	300
	5	60	600
	10	70	750
Chest (general)	<1	20	200
	5	30	400
	10	30	600
Chest (HRCT)	<1	30	50
	5	40	75
	10	50	100
Upper abdomen	<1	20	330
	5	25	360
	10	30	800
Lower abdomen & pelvis	<1	20	170
	5	25	250
	10	30	500

REFERENCES

Beaconsfield, T., Nicholson, R., Thornton, A. et al (1998) Would thyroid and breast shielding be beneficial in CT of the head? Eur. Radiol. 8, 664–667.

Calzado, A., Rodríguez, R., Muòoz, G. (2000) Quality criteria implementation for brain and lumbar spine CT examinations. Br. J. Radiol. 73: 384–395.

Calzado, A., Ruiz Sanz, S., Melchor, M. et al. (1995) A comparison of measured and calculated organ doses from CT examinations. Radiat. Prot. Dosim. 57, 381–385.

Caon, M., Bibbo, G., Pattison, J. (1999) An EGS4-ready tomographic computational model of a 14-year-old female torso for calculating organ doses from CT examinations. Phys. Med. Biol. 44: 2213–2225.

Caon, M., Bibbo, G., Pattison, J. (2000) Monte Carlo calculated effective dose to teenage girls from CT examinations. Radiat. Prot. Dosim. 90, 445–448.

CEC (1997) Directive of 30 June 1997 on Health Protection of Individuals against the Dangers of Ionising Radiation in Relation to Medical Exposures. Directive 97/43/EURATOM. Commission of the European Communities, Luxembourg.

CEC (1999) Quality Criteria for Computed Tomography, European Guidelines. EUR 16262, Commission of the European Communities, Luxembourg.

Clarke, J., Cranley, K., Robinson, J. et al. (2000) Application of draft European Commission reference levels to a regional CT dose survey. Br. J. Radiol. 73: 43–50.

Conway, B.J., McCrohan, J.L., Antonsen, R.G. et al. (1992). Average radiation dose in standard CT examination of the head: results of the 1990 NEXT survey. Radiology 184, 135–140.

Edyvean, S. (1998) Type Testing of CT Scanners: Methods and Methodology for Assessing Imaging Performance and Dosimetry. Report MDA/98/25. Medical Devices Agency, London, UK.

Geleijns, J., van Unnik, J.G., Zoetelief, J. et al. (1994) Comparison of two methods for assessing patient dose from computed tomography. Br. J. Radiol. 67: 360–365.

Gies, M., Kalender, W.A., Wolf, H. et al. (1999) Dose reduction in CT by anatomically adapted tube current modulation. I. Simulation studies. Med. Phys. 26, 2235–2247.

Greess, H., Wolf, H., Baum, U. et al. (1999) Dosisreduktion in der Computertomographie durch anatomieorientierte schwächungsbasierte Röhrenstrommodulation: Erste klinische Ergebnisse. Fortschr. Röntgenstr. 170, 246–250.

Hart, D., Hillier, M.C., Wall, B.F. et al. (1996) Doses to Patients from Medical X-Ray Examinations in the UK—1995 Review. NRPB-R289. HMSO, London, UK.

Hidajat, N., Schröder, R.J., Vogel, T. et al. (1996) The efficacy of lead shielding in patient dosage reduction in computed tomography. Röfo Fortschr. Geb Röntgenstr. Neuen Bildgeb Verfahr 165, 462–465.

Hopper, K.D., King, S.H., Lobell, M.E. et al. (1997) The breast: in-plane x-ray protection during diagnostic thoracic CT-shielding with bismuth radioprotective garments. Radiology 205, 853–858.

Huda, W., Atherton, J.V., Ware, D.E. et al (1997) An approach for the estimation of effective radiation dose at CT in pediatric patients. Radiology 203: 417–422.

ICRP (1991) 1990 Recommendations of the International Commission on Radiological Protection. ICRP Publication 60, Annals of the ICRP 21 (1–3).

ICRP (1993) Radiological protection in biomedical research. ICRP Publication 62. Annals of the ICRP 22 (3).

ICRP (1996) Radiological protection and safety in medicine. ICRP Publication 73. Annals of the ICRP 26 (2).

ICRP (2000) Pregnancy and medical radiation. ICRP Publication 84. Annals of the ICRP 30 (1).

IEC (1994) Evaluation and Routine Testing in Medical Imaging Departments, Part 2-6: Constancy Tests-X-Ray Equipment for Computed Tomography. IEC 1223-2-6. International Electrotechnical Commission, Geneva, Switzerland.

IEC (1999) Medical Electrical Equipment—Part 2: Particular Requirements for the Safety of X-Ray Equipment for Computed Tomography. IEC 60601-2-44. International Electrotechnical Commission, Geneva, Switzerland.

IPEM (1997) Recommended Standards for the Routine Performance Testing of Diagnostic X-Ray Imaging Systems. Institute of Physics and Engineering in Medicine Report No 77. IPEM, York, UK.

Itoh, S., Ikeda, M., Arahata, S. et al. (2000) Lung cancer screening: minimum tube current required for helical CT. Radiology 215, 175–183.

Jones, D.G (1997) A realistic anthropomorphic phantom for calculating organ doses arising from external photon irradiation. Radiat. Prot. Dosim. 72, 21–29.

Jones, D.G., Shrimpton, P.C. (1993) Normalised Organ Doses for X-Ray Computed Tomography Calculated Using Monte Carlo Techniques. NRPB-SR250. HMSO, London, UK.

Jurik, A.G., Bongartz, G, Golding, S.J. et al. (1998) The quality criteria for computed tomography. Radiat. Prot. Dosim. 80, 49–53.

Kalender, W.A. (2000) *Computed Tomography*. John Wiley and Sons, New York, NY.

Kalender, W.A., Schmitt, B., Zankl, M. et al. (1999) A PC program for estimating organ dose and effective dose values in computed tomography. Eur. Radiol. 9, 555–562.

Kalender, W.A., Wolf, H., Suess, C. (1999) Dose reduction in CT by anatomically adapted tube current modulation: II. Phantom measurements. Med. Phys. 26, 2248–2253.

Kalender, W.A., Wolf, H., Suess, C. et al. (1999) Dose reduction in CT by on-line tube current control: principles and validation on phantoms and cadavers. Eur. Radiol. 9, 323–328.

Kato, R., Katada, K., Anno, H. et al. (1996) Radiation dosimetry at CT fluoroscopy: physician's hand dose and development of needle holders. Radiology 201:576–578.

Keat, N.L. (2000) Real Time CT and CT Fluoroscopy. Medical Devices Agency Report MDA/00/10. HMSO, London, UK.

Lee, J.S., Primack, S.L., Staples, C.A. et al. (1994) Chronic infiltrative lung disease- comparison of diagnostic accuracies of radiography and low-and conventional dose thin-section CT. Radiology 191: 669–73.

Mayo, J.R., Harstman, T.E., Lee, K.S. et al. (1995) CT of the chest: minimal tube current required for good image quality with the least radiation dose. Am. J. Roentgenol. 164, 603–607.

Mettler, F.A., Wiest, P.W., Locken, J.A. et al. (2001) CT scanning: patterns of use and dose. J. Radiol. Prot. 20, 353–359.

Mini, R.L., Vock, P., Mury, R. et al. (1995) Radiation exposure of patients who undergo CT of the trunk. Radiology 195: 557–562.

Nagel, H.D. (2000). *Radiation Exposure in Computed Tomography* 2nd Edition. ZVEI, Frankfurt, Germany.

Naidich, D.P., Marshall, C.H., Gribbin, C. et al. (1990) Low dose CT of the lungs- preliminary observations. Radiology 175, 729–31.

Nishizawa, K., Maruyama, T., Takayama, M. et al. (1995) Estimation of effective dose from CT examination. Nippon Acta Radiologica 55: 763–768.

Olerud, H.M. (1997) Analysis of factors influencing patient doses from CT in Norway. Radiat. Prot. Dosim. 71, 123–133.

Price, R., Halson, P., Sampson, M. (1999) Dose reduction during CT scanning in an anthropomorphic phantom by the use of a male gonad shield. Br. J. Radiol. 72, 489–494.

Rehani, M.M. (2000) Computed tomography: radiation dose considerations. In: Rehani, M.M. (Ed.), *Advances in Medical Physics*. Jaypee Bros Medical Publishers, N. Delhi, India, pp. 125–133.

Rehani, M.M., Berry, M. (2000) Radiation doses in computed tomography. Br. Med. J. 320, 593–594.

Robinson, A. (1997) Radiation protection and patient doses in diagnostic radiology. In: Graigner, R.G., Allison, D.J. (Eds.), *Diagnostic Radiology: A Text Book of Medical Imaging*. Vol I. Churchill Livingstone, New York, NY, pp. 169–190.

Rogalla, P., Stover, B., Scheer, I. et al. (1999) Low-dose spiral CT: applicability to paediatric chest imaging. Pediatr. Radiol. 29, 565–569.

RCR (1998). Making the Best Use of a Department of Clinical Radiology, 4th Edition. Royal College of Radiologists, London, UK.

Rustemeyer, P., Streubuhr, U., Hohn, H.P. et al. (1999) Low-dosage dental CT. Röfo Fortschr Geb Röntgenstr. Neuen Bildgeb. Verfahr. 171, 130–35.

Seifert, H., Blass, G., Leetz, H.-K. et al. (1995) The radiation exposure of the patient from stable-xenon computed tomography. Br. J. Radiol. 68: 301–305.

Sharp, C., Shrimpton, J.A., Bury, R.F. (1998). Diagnostic Medical Exposures: Advice on Exposure to Ionising Radiation During Pregnancy. NRPB, Chilton, UK.

Shrimpton, P.C., Edyvean, S. (1998) CT scanner dosimetry. Br. J. Radiol. 71, 1–3.

Shrimpton, P.C., Jessen, K.A-, Geleijns, J. et al. (1998) Reference doses in computed tomography. Radiat. Prot. Dosim. 80, 55–59.

Shrimpton, P.C., Jones, D.G., Hillier, M.C. et al. (1991) Survey of CT practice in the UK. Part 2: Dosimetric aspects. NRPB-R249. HMSO, London, UK.

Shrimpton, P.C., Wall, B.F. (1995) The increasing importance of x-ray computed tomography as a source of medical exposure. Radiat. Prot. Dosim. 57, 413–415.

Shrimpton, P.C., Wall, B.F. (2000) Reference doses for paediatric CT. Radiat. Prot. Dosim. 90, 249–252.

UNSCEAR (2000) United Nations Scientific Committee on the Effects of Atomic Radiation 2000 Report to the General Assembly, Annex D: Medical Radiation Exposures. United Nations, New York, NY.

Veit, R., Zankl, M., Petoussi, N. et al. (1989) Tomographic anthropomorphic models. Part 1: construction technique and description of models of an 8 week old baby and a 7 year old child. GSF-Bericht 3/89. Gesellschaft für Strahlenforschung, Neuherberg, Germany.

Wall, B.F., Hart, D. (1997) Revised radiation doses for typical x-ray examinations, report on a recent review of doses to patients from medical x-ray examinations in the UK by NRBP. Br. J. Radiol. 70, 437–439.

Wall, B.F., Shrimpton, P.C. (1998) The historical development of reference doses in diagnostic radiology. Radiat. Prot. Dosim. 80, 15–20.

Weisser, G., Lehmann, K.J., Scheck, R. et al. (1999) Dose and image quality of electron-beam CT compared with spiral CT. Invest. Radiol. 34, 415–420.

Xu, X.G., Chao, T.C., Bozkurt A. (2000) VIP-MAN: an image-based whole-body adult male model constructed from colour photographs of the visible human project for multi-particle Monte Carlo calculations. Health Phys. 78, 476–486.

Zankl, M. (1998) Methods for assessing organ doses using computational models. Radiat. Prot. Dosim. 80, 207–212.

Zankl, M., Panzer, W., Drexler, G. (1991) The calculation of dose from external photon exposures using reference human phantoms and monte carlo methods. part vi: organ doses from tomographic examinations. GSF-Bericht 30/91. Gesellschaft für Strahlenforschung, Neuherberg, Germany.

Zankl, M., Panzer, W., Drexler, G. (1993) Tomographic anthropomorphic models. Part II: organ doses from computed tomographic examinations in paediatric radiology. GSF-Bericht 30/93. Gesellschaft für Strahlenforschung, Neuherberg, Germany.

Zankl, M., Panzer, W., Petoussi-Henss, N. et al. (1995) Organ doses for children from computed tomographic examinations. Radiat. Prot. Dosim. 57, 393–396.

In compliance with GPSR, should you have any concern about the safety of this product, please advise: International Associates Auditing and Certification Limited, The Black Church, St Mary's Place, Dublin 7, D07 P4AX Ireland EUAR@ie.ia-net.com

Printed and bound by CPI Group (UK) Ltd, Croydon, CR0 4YY

14/09/2026

02216204-0001